STORY MONSTER'S

S.T.E.A.M. CHRONICLES

BOOK ONE

Night Watchers

NOCTURNAL CREATURES OF NORTH AMERICA

COYOTES, SNAKES, OWLS, SPIDERS, AND MORE

WRITTEN BY
CONRAD J. STORAD

ILLUSTRATED BY
JEFF YESH

STORY MONSTERS® PRESS
An imprint of Story Monsters LLC
Chandler, Arizona, USA

Linda F. Radke, Publisher
Story Monsters Press
An imprint of Story Monsters LLC
4696 W. Tyson Street
Chandler, AZ 85226
480-940-8182
Publisher@storymonsters.com
www.StoryMonstersPress.com

STORY MONSTERS® PRESS

Publisher's Cataloging-in-Publication

Names: Storad, Conrad J., author. | Yesh, Jeff, 1971- illustrator.

Title: Story Monster's S.T.E.A.M. chronicles. Book one, Night watchers : nocutural creatures of North America : coyotes, snakes, owls, spiders, and more / written by Conrad J. Storad ; illustrated by Jeff Yesh.

Other titles: Night watchers : nocutural creatures of North America : coyotes, snakes, owls, spiders, and more

Description: Chandler, Arizona, USA : Story Monsters Press, an imprint of Story Monsters LLC, [2024] | Interest age level: 8 and up. | Summary: Readers join Story Monster as Eddie the Elf Owl serves as a learning guide to explore the amazing world of nocturnal creatures that live throughout North America.--Publisher.

Identifiers: ISBN: 978-1-58985-276-1 (paperback) | 978-1-58985-278-5 (hardcover) | 978-1-58985-277-8 (ebook)

Subjects: LCSH: Monsters--Juvenile fiction. | Owls--Juvenile fiction. | Nocturnal animals--Juvenile fiction. | Animals--North America--Juvenile fiction. | Insects--North America--Juvenile fiction. | CYAC: Monsters--Fiction. | Owls--Fiction. | Nocturnal animals--Fiction. | Animals--North America--Fiction. | Insects--North America--Fiction.

Classification: LCC: PZ10.3.S891 Ni 2024 | DDC: [Fic]-dc23

Includes bibliographical references and index.
ISBN-10: (paperback)1-58985-276-1
ISBN-13: (paperback) 978-1-58985-276-1
ISBN-10: {hardcover} 1-58985-278-8
ISBN-13: {hardcover} 978-1-58985-278-5
Ebook ISBN-10: 1-58985-277-X
Ebook ISBN-13: 978-1-58985-277-8

Printed in the United States of America
Author: Conrad J. Storad
Cover design, interior design, and illustrations: Jeff Yesh
Proofreader: Cristy Bertini
Project Manager: Linda F. Radke

To Laurie, Sarah, Meghan, Natalie, and Hadley Jane. You brighten my life every single day!

– Conrad J. Storad

ONe HOT NiGHT iN a WesTeRN DeseRT, MaNy yeaRS aGO ...

A huge silver moon rose above the distant mountains. After a blistering hot day, the air was finally a bit cooler. Stars filled the sky.

Tall, spiny plants were silhouetted against the horizon. They looked like giants with many arms. The little monster looked up at the night sky, its four eyes open wide. It blinked. The moonlight made its lime green skin appear to glow.

A voice startled the monster.

"Whooo. Whooo. Whooo are you?"

There was a hole high in the trunk of one of the tall plants. It was a giant cactus called a saguaro. The voice was coming from there.

"Whooo are you? My name is Eduardo E. Owlington VIII," said a small creature peeking out of the hole.

"My friends just call me Eddie. I'm an Elf Owl. I'm a night watcher. My job is to watch this small part of the desert. My family has had this job for eight generations."

The monster said nothing. It shimmered with a purple glow. Slowly, it changed shape until it looked exactly like the elf owl in the cactus above.

Eddie was stunned by what he saw. Then he smiled. His yellow eyes seemed to sparkle.

"Oh my. I know exactly who and what you are.

My great-grandfather told me about your kind," said Eddie.

"You must be a Story Monster. You're a shapeshifter.

You like to watch and learn, don't you?"

There was another shimmer of purple. The little monster appeared to shift back and forth from the shape of an owl to its own four-eyed form. The long eye stalks seemed to nod Yes to Eddie's questions.

"Most excellent!" said Eddie.
"Let me tell you all about my family and about some of the other critters of the night."

Eddie was a talker. His talk became a learning experience.

Owls are known as nocturnal birds. But there are many other kinds of nocturnal creatures as well. Not all of them are birds. There are nocturnal mammals and reptiles and insects.

Nocturnal simply means that an animal is most active during dark hours. They rest during the day when the sun is bright and hot.

"Diurnal creatures are different," Eddie said. "They sleep at night and work during the day. Well, except for those humans who work third shift," he giggled. "We call them night owls."

The Story Monster blinked its four eyes, one by one by one by one. It just stared at Eddie, listening to his every word.

Eddie just kept talking. He hardly took a breath.

"I'm an Elf Owl, the smallest owl living in the Southwest deserts. I'm also very smart," Eddie added, without cracking a grin.

There are many kinds of owls. Eddie's cousins include Pygmy owls, Burrowing owls, Barred owls, Barn owls, and the Snowy owls. All are night watchers. And all are excellent hunters.

BARN OWL

BARRED OWL

SNOWY OWL

Eddie's voice took on a serious tone.

"The biggest and baddest of my cousins are the Great Horned Owls.

They live all over North and South America," Eddie said with pride.

Eddie quickly added, "Don't mess with them, little monster!"

Some people call them tiger owls or hoot owls. There are lots of stories about them. Great Horned Owls are fierce predators. They are not afraid of anything, including other raptors.

Like most owls, Great Horned Owls have huge eyes and excellent night vision, even though the eyeballs don't move in their sockets. They can swivel their entire head to look in any direction.

The Story Monster's head began to rotate almost in a full circle.

GREAT HORNED OWL

Eddie loved to talk. He was on a roll.

"Humans are funny. Of course, I call them 2-leggers," Eddie chuckled.

Then he told the little monster something interesting about humans. Eddie said that lots of them are afraid of monsters. The Story Monster was smart to stay out of sight in the shadows.

"Some humans also get scared of things that fly at night," Eddie said. That's why they don't like my friends, the bats."

As Eddie talked, the little monster seemed to grow a pair of thin, leathery wings.

Bats are amazing nocturnal creatures. In America, they live in all 50 states. And remember, bats are NOT birds. They're mammals. In fact, bats are the only mammal that can fly... without an engine that is.

"Some bats eat fruit and some bats like fish. But, like me, most bats prefer the bug buffet," Eddie said.

Eddie listed some fun facts. One hungry little brown bat can gobble as many as 1,000 mosquitoes in an hour. Other kinds of bats can eat their body weight in insects every night.

1,000
MOSQUITOES
IN 1 HOUR

The Story Monster's belly bulged and got rounder.

Eddie finally took a breath. Then he talked more about bats.

"Some very interesting bats live here in the Sonoran Desert. They're good friends who do important work," Eddie explained.

LONG-NOSED BAT

Long-nosed bats eat more than just moths and bugs. They also like nectar and sticky pollen. They are pollinators. They can hover like a hummingbird in front of a blooming saguaro flower to eat their fill. Without them, the giant cactus Eddie lived in would never have grown.

Story Monster morphed from a bat to a giant cactus, and back.

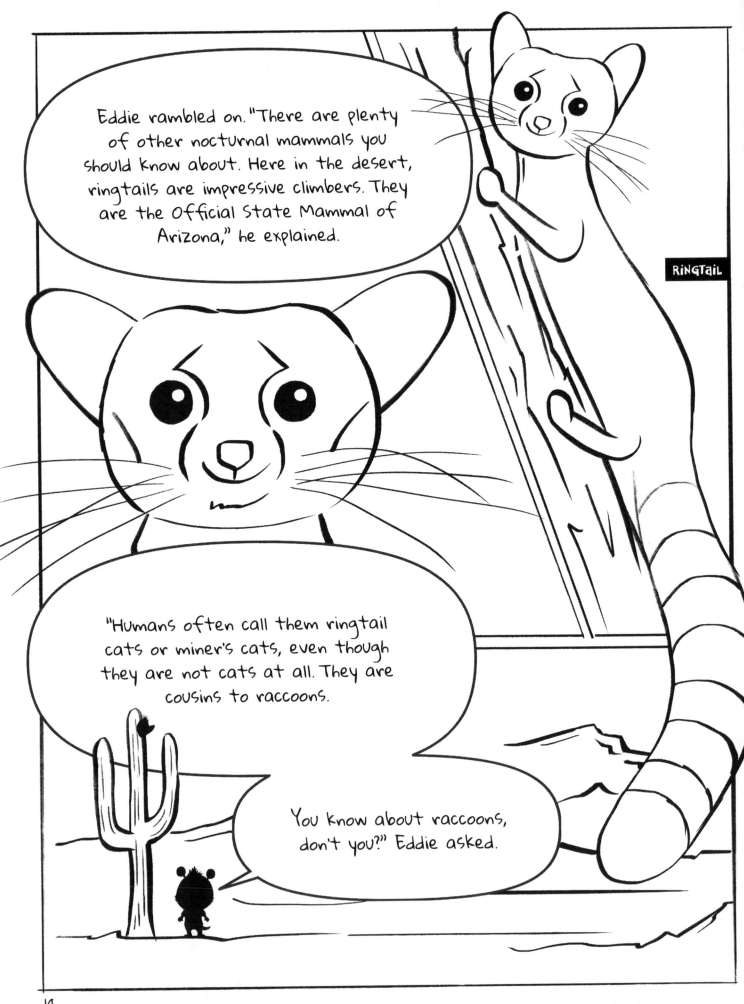

Eddie rambled on. "There are plenty of other nocturnal mammals you should know about. Here in the desert, ringtails are impressive climbers. They are the Official State Mammal of Arizona," he explained.

RINGTAIL

"Humans often call them ringtail cats or miner's cats, even though they are not cats at all. They are cousins to raccoons.

You know about raccoons, don't you?" Eddie asked.

14

Raccoons live all over North America and in other parts of the world. They're smart and crafty. They're also excellent swimmers.

RACCOON

"Raccoons love the woods, but they also know how to survive in big city neighborhoods. They'll eat almost anything," Eddie laughed. "Humans need to secure their garbage cans."

The little monster seemed to grow a bushy black and white striped tail and a black mask covered two of its eyes.

Rocky's DINER

Eddie was enjoying his talk with the Story Monster, even though the little green creature hadn't said a word. Eddie just kept spewing out information.

"Let me tell you about the opossum," Eddie said." The opossum is the only marsupial that lives in North America. It's a "pouched mammal."

Humans know a bit more about the opossum's more famous cousins, the kangaroos and koalas that live in Australia. Guess what, opossums have lived on this planet for more than 70 million years.

KANGAROO

KOALA

70 MILLION YEARS

The little monster bounced like a kangaroo, for just a second.

"Opossums are not handsome, like me," Eddie went on.

"They kind of look like a big rat with a long tail. But that tail is special."

Opossums use their tails to grab onto branches and for balance as they move around in trees.

OPOSSUM

"You should know they have 50 little sharp teeth. That's more than any land mammal in North America," Eddie added.

Opossums do good work with those teeth. They help keep neighborhoods clean by eating lots of garden pests and rodents. They even eat ticks and cockroaches. Yuck!

As Eddie talked, the Story Monster shimmered and flickered. For a moment it looked like a huge wolf.

Then it appeared to be a little girl with a red cape carrying a picnic basket filled with goodies.

Eddie was startled. He laughed, then just kept talking.

"Oh yes. Wolves," he said. "They are nocturnal mammals. There are timber wolves, gray wolves, and red wolves in North America. Just not as many as there used to be. But the Mexican Gray Wolf is making a comeback in Arizona and other parts of the Southwest."

Eddie finally took a breath. But the little monster had changed shape again. Now it looked like a howling coyote.

"Of course! Coyotes," Eddie laughed. "There are more coyotes than ever before. You can find them everywhere in North America

They love to sing. Native people called them Song Dogs. They use at least 10 different sounds to communicate. They bark, yip, growl, and howl. No texting for coyotes."

HA HA

coyote

The Story Monster flickered back to its four-eyed form.

Eddie was like a feathered encyclopedia. The little owl was bursting with information. He knew all kinds of fun facts.

Red Fox

He said the red fox is a wily hunter, and a talker, too. They have about 20 different calls. They're also one of the most widespread meat-eating mammals on our planet.

"A red fox is not a picky eater," Eddie explained. "They'll eat birds, frogs, snakes, bugs, and even berries."

The monster was shimmering again. It seemed to grow a long fuzzy striped tail in the shape of a question mark. Then it changed again. Now it looked kind of like a pig with a long snout. But it was covered with a bony shell made of nine bands. The skinny tail was long and segmented. Eddie watched in amazement.

The Story Monster just stared. Eddie didn't waste a breath.

HA
HA

"Little monster, you know more than I thought," he laughed.

"Yes, indeed. Skunks and armadillos are nocturnal mammals."

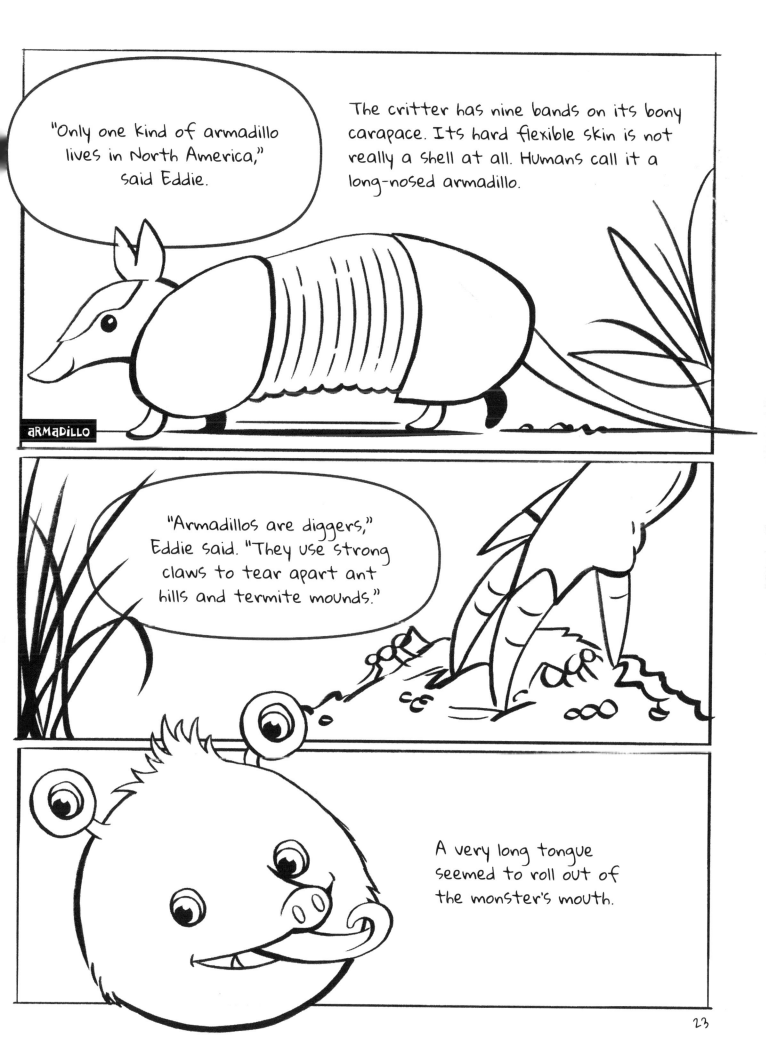

"Only one kind of armadillo lives in North America," said Eddie.

The critter has nine bands on its bony carapace. Its hard flexible skin is not really a shell at all. Humans call it a long-nosed armadillo.

ARMADILLO

"Armadillos are diggers," Eddie said. "They use strong claws to tear apart ant hills and termite mounds."

A very long tongue seemed to roll out of the monster's mouth.

The Story Monster was wide-eyed again.

The owl explained about four different kinds of skunks that live in North America. There are hooded, hog-nosed, striped, and spotted skunks. All are omnivores. That means they eat both plants and animals, mostly insects and grubs.

"A word of warning little fellow," Eddie continued. "Don't mess around with skunks. Never, ever corner one or threaten its babies."

HOODED SKUNK

HOG-NOSED SKUNK

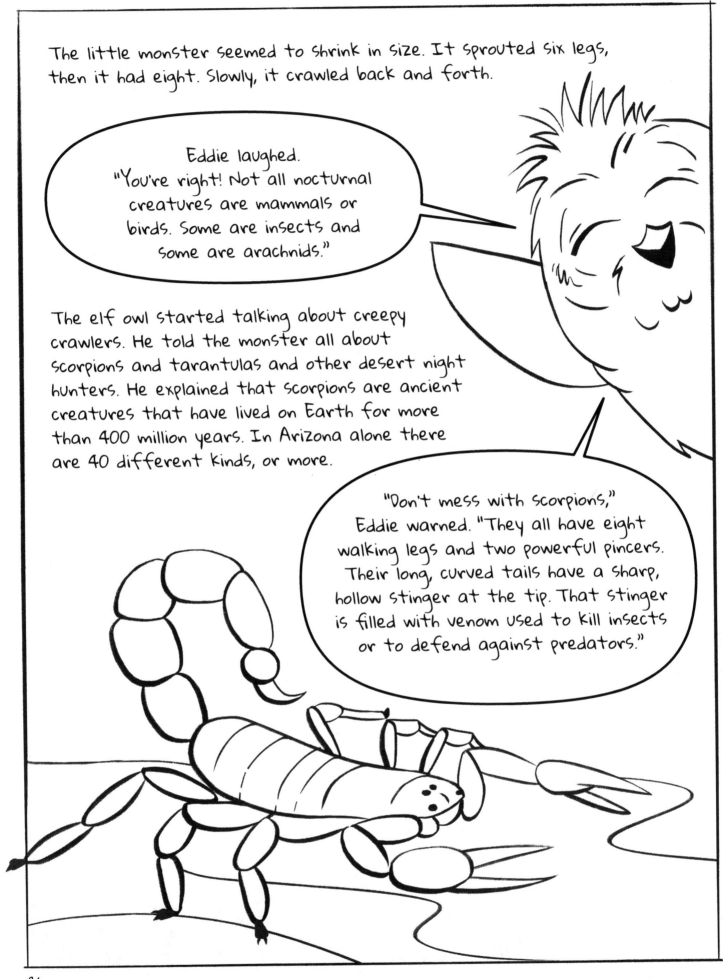

The little monster seemed to shrink in size. It sprouted six legs, then it had eight. Slowly, it crawled back and forth.

Eddie laughed. "You're right! Not all nocturnal creatures are mammals or birds. Some are insects and some are arachnids."

The elf owl started talking about creepy crawlers. He told the monster all about scorpions and tarantulas and other desert night hunters. He explained that scorpions are ancient creatures that have lived on Earth for more than 400 million years. In Arizona alone there are 40 different kinds, or more.

"Don't mess with scorpions," Eddie warned. "They all have eight walking legs and two powerful pincers. Their long, curved tails have a sharp, hollow stinger at the tip. That stinger is filled with venom used to kill insects or to defend against predators."

The Story Monster was still. It now had eight legs and a hairy body. Eddie noticed the change, but just kept right on talking.

He explained that spiders and scorpions were cousins. He said that there were thousands of kinds of spiders living on the Earth. More than 4,000 kinds in North America alone.

4,000
KINDS OF SPIDERS
IN NORTH AMERICA

TARANTULA

"Tarantulas are among the biggest spiders around here," Eddie said.

"Most of them are gentle giants. Still, I'd play it safe and keep my distance, little guy."

The Story Monster seemed to enjoy shape shifting. The eight legs changed to six. The body grew longer and sprouted two antennae and wings. Then the bottom of its body glowed with a soft yellow light.

"Yes, yes, of course!" Eddie almost shouted. "Many types of insects are nocturnal. And the firefly is among the most famous."

FIREFLY

He explained that crickets and cicadas and katydids are singers. Mosquitoes are little vampires. And that "lightning bug" was another human name for the firefly.

"But guess what, little monster?" Eddie laughed. "The firefly is not a fly or a bug at all. It's a flying beetle."

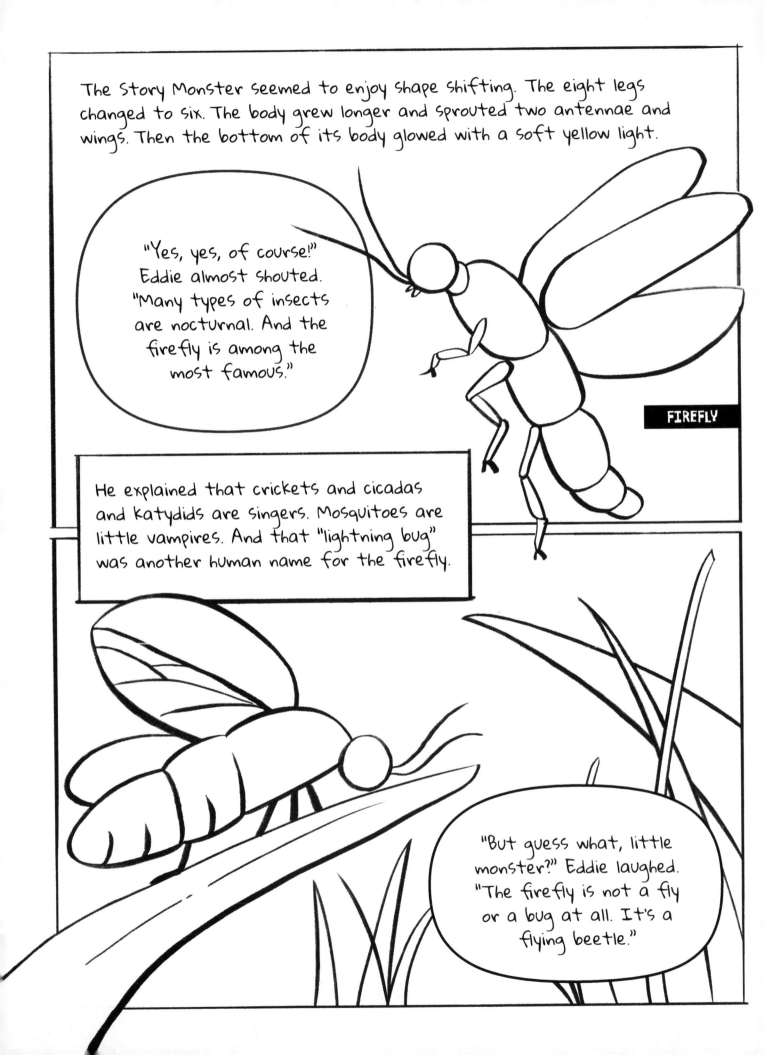

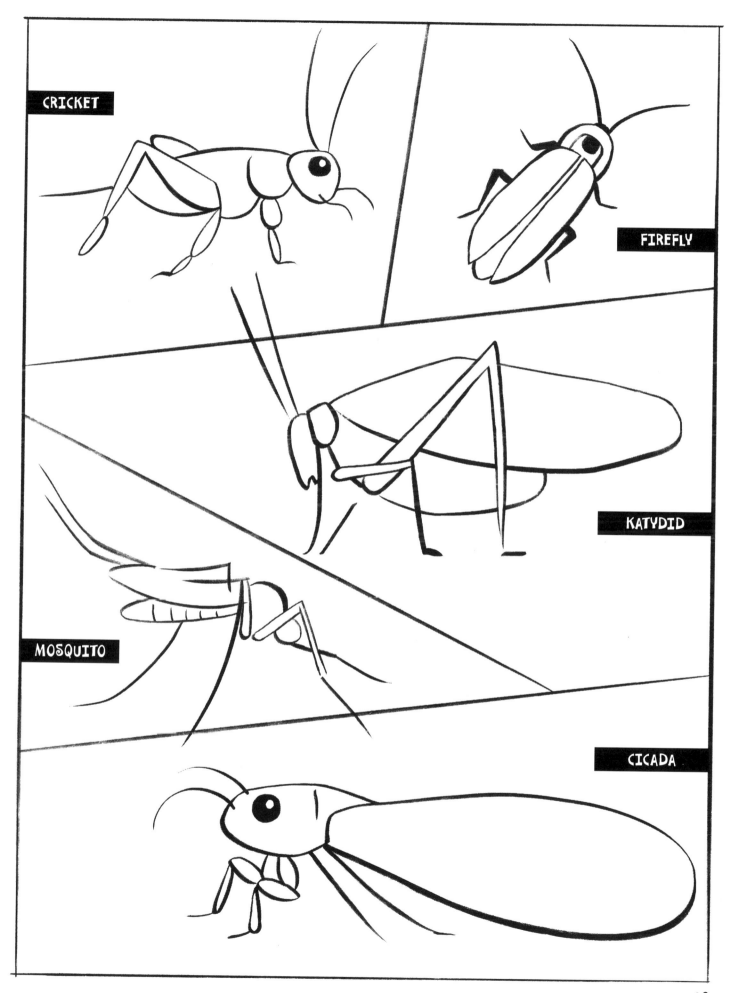

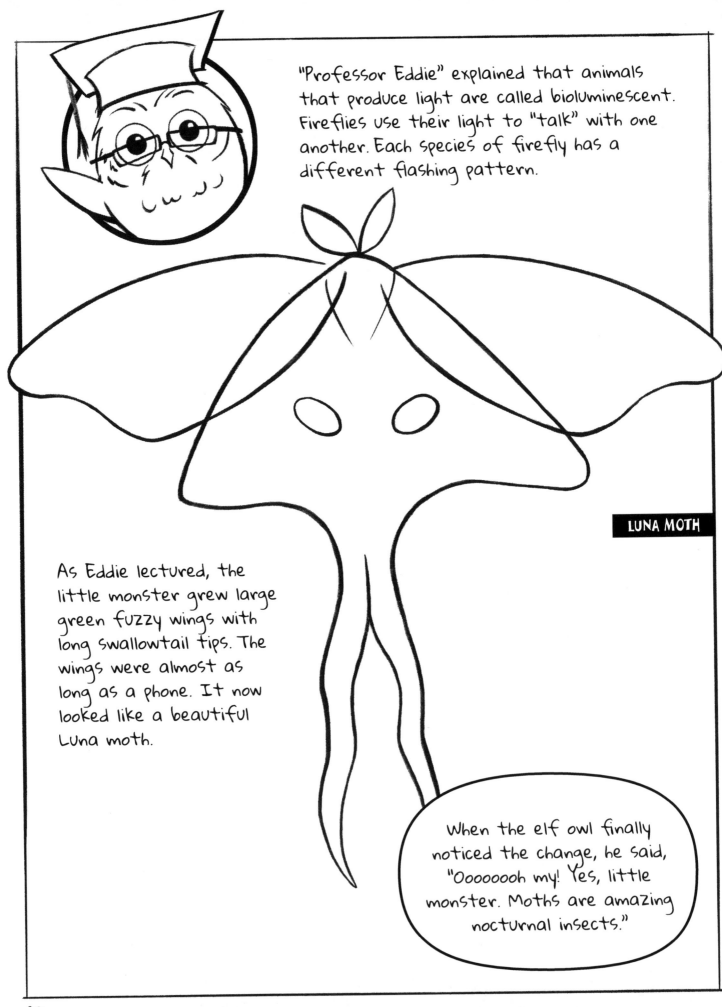

"Professor Eddie" explained that animals that produce light are called bioluminescent. Fireflies use their light to "talk" with one another. Each species of firefly has a different flashing pattern.

LUNA MOTH

As Eddie lectured, the little monster grew large green fuzzy wings with long swallowtail tips. The wings were almost as long as a phone. It now looked like a beautiful Luna moth.

When the elf owl finally noticed the change, he said, "Oooooooh my! Yes, little monster. Moths are amazing nocturnal insects."

Eddie launched into a list of facts about moths. He said that moths were some of the most beautiful of all nocturnal creatures. But that most humans never saw or noticed them.

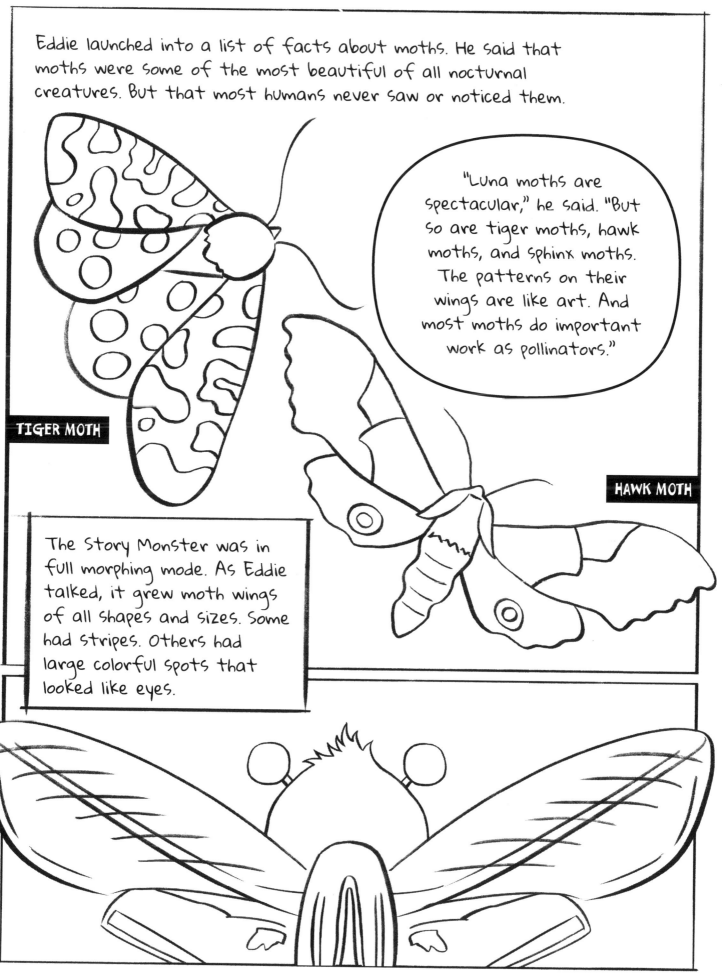

"Luna moths are spectacular," he said. "But so are tiger moths, hawk moths, and sphinx moths. The patterns on their wings are like art. And most moths do important work as pollinators."

TIGER MOTH

HAWK MOTH

The Story Monster was in full morphing mode. As Eddie talked, it grew moth wings of all shapes and sizes. Some had stripes. Others had large colorful spots that looked like eyes.

The bright moon was dipping toward the western horizon. Eddie had been talking for hours. He didn't sound tired at all.

The little monster had lost the colorful wings. It had shrunk into the shape of a tiny toad. Or was it a frog? Its throat swelled like a large chewing gum bubble and looked ready to pop.

The elf owl took a breath. He saw the bubble and started to laugh. Then he made a "Ribbit, ribbit!" sound.

"RiBBiT, RiBBiT!"

"You are so smart, little monster," Eddie said. He hadn't talked at all about nocturnal reptiles or amphibians.

TOAD

Many snakes are night hunters. But his favorite were the frogs and toads.

32

Eddie looked down from his perch to see if the Story Monster had changed shapes. All that remained was a thin cloud of purple mist that quickly drifted away.

The elf owl chuckled to himself.

"I always talk too much."

Off in the distance, high atop a big cactus,
Eddie thought he saw a shimmering purple shape.
It looked kind of like another elf owl.

"Come back any time little Story
Monster," Eddie called out.

"I have lots more
stories to tell."

But the shape was gone.

Activity Guide

We love stories and books. To make this one a favorite, let's explore the story through seeing, reading, writing, listening, speaking, and touching.

1. How many words can you make out of the word N O C T U R N A L ?

_____ _____ _____ _____

_____ _____ _____ _____

2. Compare *Night* noises versus *Day* noises.

   a. Stand outside your house during the day for about 10 minutes.
   What are the sounds that you hear? Write them down.

   _____

   b. Stand outside your house during the night for about 10 minutes.
   What sounds do you hear? Write them down.

   _____

   c. How do the noises differ? How are the noises similar?

   _____

If you were outside at night in the country you could hear nocturnal animals.
Listen:

- Owl sounds: https://abcbirds.org/blog21/owl-sounds/

- Coyote yelps and reasons why: https://coyoteyipps.com/coyote-voicings/

- Toad sounds: https://www.nps.gov/subjects/sound/cspadefoottoad.htm

- Frog sounds: https://thefroglady.wordpress.com/2019/03/26/frogs-of-arizona/

- Bat sounds: https://www.nps.gov/subjects/bats/echolocation.htm

> Try to imitate the sounds with your voice - it's fun!

3. Flashlight Outlines

Photocopy and cut out pictures of animals from this book.

a. Make a shadow show: Cut a small slit in a plastic straw and insert each animal.

b. In a dark room or at night outside, shine a flashlight against a solid wall and then hold and twist the straw in front of the light to make the animals dance and spin on the wall.

4. Story Time!

Tell me a story about two night animals meeting each other in the moonlight. Here are the scenes:

a. The centipede is going along and the coyote almost steps on him. What could be their conversation?

b. A fox and a skunk are both trying to hide behind the same cactus plant. Can you tell the story about this encounter?

5. Give names to three nocturnal critters.

_____

Why did you choose those names?

_____

6. Some nocturnal creatures crawl, some slither, some hop, some fly, some run, and some walk. What a lot of differences! Name at least one creature for every type of locomotion?

_____

_____

7. Vocabulary Match

a. Match these Vocabulary Words in the book with their correct definition? Put the correct letter of the definition under the Matching Definition column.

b. Give an example of an animal from the book that matches the definition? Write the animal under the Animal Example column.

Vocabulary Match

Animal Example	Vocabulary Word	Matching definition	Definitions
	1. Raptor		**A.** Take pollen from one plant to another
	2. Nocturnal		**B.** Babies live in a pouch on the mother
	3. Diurnal		**C.** Disappear and appear in another place
	4. Predator		**D.** A lot of choices of food to eat
	5. Shapeshifter		**E.** Hunt mostly in the daytime
	6. Mammals		**F.** A story that isn't true
	7. Buffet		**G.** Glow or blink a light in the dark
	8. Pollinator		**H.** Crafty and clever
	9. Morphed		**I.** Able to change to a different appearance
	10. Myth		**J.** Animals that hunt other animals
	11. Colony		**K.** Flying birds that capture food with their claws
	12. Marsupial		**L.** A group of same animals
	13. Wily		**M.** Animals that give birth to live young
	14. Bioluminescent		**N.** Hunt mostly in the nighttime
			Answers are on the next page

8. Here are some unique features. Name the animals that have these abilities:

a. Can turn its head all the way around.

_____

b. Uses a foul smell for defense.

_____

c. Protected by a hard shell or hard skin.

_____

d. Glows in the dark.

_____

9. Let's do some research.

There are many different species of nocturnal animals.
Find the number of species for each of these.

a. Fox

b. Bat

c. Owl

d. Skunk

10. Now it's time to use your creative imagination.

Which animal are you?

What adventure will you go on tonight?

Answer to vocabulary match:

1.K, 2.N, 3.E, 4.J, 5.I, 6.M, 7.D, 8.A, 9.C, 10.F, 11.L, 12.B, 13.H, 14.G

The Creative Team

Conrad Storad – Author

Conrad J. Storad is the award-winning author and editor of more than 60 science and nature books for young readers. He also writes a monthly science/nature column called "Conrad's Classroom" in *Story Monsters Ink*, a national award-winning magazine. Many of Storad's books reflect his interest and fascination with nature's amazing plants and creatures. An accomplished storyteller, Storad has "edu-tained" more than 1 million students and teachers across the nation with his programs and writing workshops that promote reading and science literacy. After 33 years living and working in Tempe, Arizona, Conrad and wife Laurie now live in their hometown of Barberton, Ohio, where he was a 2017 honoree on the city's Walk of Fame.

Jeff Yesh – Illustrator

Jeff Yesh is a freelance illustrator and graphic designer whose award-winning work has been featured in multiple children's books. Yesh is also the artistic talent behind the Story Monster character and is the graphic designer for the award-winning *Story Monsters Ink*, the literary resource for parents, teachers, and librarians. Yesh graduated from Indiana State University with a Bachelor of Fine Arts in Graphic Design. He lives in Indiana with his wife, two daughters, and a slew of pets.

STORY MONSTERS' PRESS

An imprint of Story Monsters LLC
Chandler, Arizona, USA

Story Monsters Press – Publisher

Story Monsters Press, an imprint of Story Monsters LLC, is a publisher of children's books that offer hope, value differences, and build character. Each book also includes a curriculum guide complementing the story for parents and educators to use with young readers.

Made in the USA
Columbia, SC
28 September 2024

43055004R00024